The Geometry
of Light

Galileo's Telescope, Kepler's Optics

Gerald Rottman

Published by Gerald Rottman, Baltimore, Maryland.
www.TheGeometryOfLight.com

Publisher's Cataloging-in-Publication Data

Rottman, Gerald.
The geometry of light : Galileo's telescope, Kepler's optics / Gerald Rottman.
   p. cm.
   Includes bibliographical references and index.
   ISBN 978-0-9819416-0-8

1. Geometrical optics. 2. Optics –History. 3. Optical instruments. 4. Geometrical
optics–History. 5. Astronomical instruments –Europe –History –17th century.
I. Title.

QC381 .R68 2008
535/.32–dc22   2008910726

Printed in the United States of America.

10 9 8 7 6 5 4 3 2 1

Preface

Johannes Kepler was a giant of seventeenth-century science. A contemporary of Galileo, Kepler is principally known as a founder of modern astronomy. But Kepler was also active in the field of optics. Inspired by Galileo's discoveries of 1609 and 1610, Kepler developed a theory of lenses to explain the operation of Galileo's telescope. Kepler published his pioneering work in 1611 as a short book titled *Dioptrice*. The Latin title (pronounced di-óp-tri-ke), is roughly translated as "the study of refraction."

The Geometry of Light: Galileo's Telescope, Kepler's Optics presents the main ideas and methods of *Dioptrice*. Why, four hundred years after its publication, should Kepler's book interest modern readers? First, *Dioptrice* deals with questions that almost everybody has wondered about, how vision occurs and how lenses work. Second, Kepler's geometric approach conveys an intuitive grasp of optics that is hard to obtain from modern methods. Finally, Kepler's theory of lenses has a special charm because it achieves so much with so little. It is truly a breathtaking experience to follow Kepler as he deftly lays the foundations of modern optics using only a few simple principles.

Dioptrice is not an easy read, and no English translation of the original Latin work is currently available. *The Geometry of Light* presents the main ideas of *Dioptrice* in a way that I hope will be accessible to a general audience. In this volume I provide a fairly complete account of Kepler's concepts and mode of reasoning. But I have been selective regarding applications of the theory. In the main, only those discussions that build toward an understanding of the telescope are presented. At the same time, I often say more than Kepler does. Kepler does not always explain or justify his assertions fully. In those instances, I have provided more complete explanations.

No familiarity with optics or the history of optics is assumed. Chapter 1 provides historical and conceptual background, while the appendices review the needed mathematics. Chapter 2 introduces Kepler's conceptual tools, highlighting the simple rule of refraction that he uses so effectively. Chapter 3 accounts for the properties of convex lenses using the concept of a convergence point. This is followed by a discussion of image formation on the retina by the lens of the eye.

Chapter 4 explains the sometimes puzzling experience of viewing objects through a convex lens. This lays the groundwork for explaining the appearance of objects through combinations of lenses. The final chapter presents Kepler's original design for a telescope using two convex lenses. Galileo's telescope, however, used one concave and one convex lens. So the discussion turns to the concave lens, and this leads to Kepler's explanation of the Galilean telescope. The book concludes with a brief note about the sine law of refraction.

I have attempted to present the material in a way that will be interesting to those who are knowledgeable about optics, yet accessible to those who are not. The only prerequisites are the patience to examine and think about the diagrams that are provided, and a reasonable comfort with high school mathematics. This volume may also be useful as enrichment material for motivated high school students during their first encounters with geometry or physics. Issues significant to a historian of science are not addressed, but this volume may be a useful starting point for such inquiries.

It is a pleasure to acknowledge the dedicated and expert editorial support of my daughter Ayda. Finally, I wish to thank my beloved wife, Elka, to whom I owe the peace of mind that has enabled me to write this book.

Gerald Rottman
Baltimore, 2008

Contents

Chapter 1

Historical and conceptual background

Galileo's telescope was a simple instrument consisting of just two lenses. The subtlety of its construction lay in choosing lenses of the proper shape, separating the lenses by the proper distance, and using lenses of sufficiently high quality. Galileo constructed his telescope on the basis of practical experience he had gained from handling lenses.

Galileo proceeded to use his telescope to make far-reaching discoveries in astronomy that challenged the geocentric model of the solar system. Kepler's work in theoretical astronomy complemented Galileo's observations in promoting the heliocentric model of the solar system. Galileo and Kepler also had complementary relationships to the telescope. While Galileo continued his observations, Kepler devoted himself to theoretical work on the optics of the telescope.

Kepler perceived that the telescope produced its effect by means of the refraction of light. He wanted to understand in detail how the refraction of light could result in the magnification of distant objects. In a short period of intense work, Kepler developed a successful theory of lenses and telescope design based on just a few simple principles.

Kepler recorded his work on lenses in a short book titled *Dioptrice*. Previously, he had written a larger work on optics, *Paralipomena to Witelo and the Optical Part of Astronomy*. There, Kepler dealt with the nature of light and vision as they relate to astronomy. This chapter reviews what is needed from Kepler's first book on optics, as well as other background necessary to understand his theory of lenses.

Euclidean optics

Kepler's first book on optics dealt, among other topics, with the question of how it is possible for the eye to perceive distant objects. Logically, some means of bridging the space that separates object and observer is required. In analogy to the sense of touch, one approach would be to postulate that the eye in some way reaches out to the object. Another possibility would be to postulate that the object transmits something to the eye.

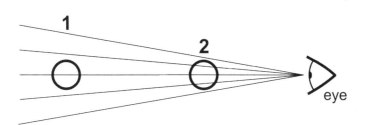

Figure 1.1 Rays emerging from the eye.

This question goes back at least as far as Euclid, who lived some time around 300 BC. Of course, Euclid is the author of the *Elements*, the famous compendium of geometry. But Euclid also investigated optics, using geometry as his principal tool. Euclid held that *rays* extend as straight lines from the eye, acting like probes to detect what lies in front of the eye. As Figure 1.1 shows, Euclid imagined a cone of rays with its apex at the eye. The observer sees a given object only if that object intersects one or more of these rays.

Euclid also postulated that the apparent size of an object depends on how many rays it intersects. The farther away the object is, the fewer rays it intersects and the smaller it appears. Thus, in Figure 1.1 the same object appears smaller when at position 1 than when at position 2.

Arabic advances in optics

Al-Kindi was an Arabic scholar who lived in Baghdad during the ninth century. He promoted the study of classical Greek sources. In a reversal of Euclid's scheme, Al-Kindi thought in terms of rays emerging from each point on the surface of a visible object, rather than from the eye (Figure 1.2). Complex visual scenes are thus conceived as assemblies of points, each point seen independently of any other when a ray from that point enters the eye.

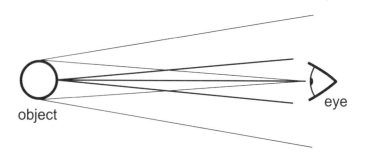

Figure 1.2 Rays diverging from representative points of an object.

Another important Arabic scholar was Alhazen, who lived in Cairo during the eleventh century. He arrived at the current understanding that vision results from the eye's sensitivity to light. This stood in opposition to Aristotle's conception of light. For Aristotle, light was an adjective, not a noun. Aristotle conceived of light as the state of transparency of the medium that fills space. When the medium is in its "light state," it is transparent. When in its "dark state," the medium is not transparent. Breaking with Aristotle and developing Al-Kindi's ideas, Alhazen held that light is a thing that moves through space. When light enters the eye from any point on the surface of an object, the observer sees that point.

This explanation is not without its problems. If rays of light arrive at the eye from innumerable points on the surface of an object, how does the eye sort out these rays to form an impression of the object? A hint to how this might occur is given by the *camera obscura*.

The camera obscura

A camera obscura is a darkened room into which light from outside is admitted through a narrow opening, or *aperture*, in one wall, and an image of the scenery outside is formed on the opposite wall. The modern device used to take photographs is called a *camera* because of its similarity to the camera obscura.

The formation of an image by the camera obscura can be explained on the basis of two simple assumptions:

- Light diverges in all directions from any illuminated point.

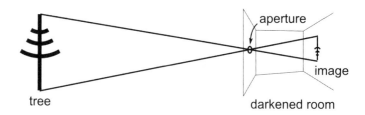

Figure 1.3 Image formation in a camera obscura.

- The path that light takes when it moves from one place to another is a straight line.

For instance, consider how a camera obscura forms the image of a tree. Only a tiny fraction of the light emerging from any point of the tree travels in the right direction to enter the aperture of the camera obscura. Figure 1.3 illustrates the path along which light from the top of the tree passes through the aperture to illuminate a small spot on the wall. Because light from the top of the tree travels along a downward-sloping line through the aperture, this spot will be lower than the aperture.

Similarly, light from a point at the base of the tree travels along an upward sloping line, so the spot illuminated by this point will be higher than the aperture. Continuing in this fashion, it is not difficult to see that light from all the points of the tree will be distributed on the wall in a pattern that reproduces the arrangement of the source points, only inverted. The resulting image of the tree will be upside-down and backward.

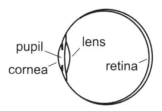

Figure 1.4 Structure of the eye.

The camera obscura illustrates how an image can be formed merely because light travels along straight lines. But it is critical that the aperture be narrow so that light from any point illuminates only a point-like area of the wall. With a large aperture, light from different points would illuminate large overlapping areas of the wall and no image would be apparent.

This implies that the eye must form images by a different means. The pupil of the eye is a narrow opening only when constricted in bright light. In dim light, it dilates to a size comparable to the diameter of the eye itself, yet the eye still sees (Figure 1.4).

Kepler believed that the eye projects an image on the retina at the back of the eye in a manner that superficially resembles the camera obscura. Unlike the camera obscura, however, the eye has a lens behind its opening. Kepler had the intuition that this lens would suffice to direct light from different points of visible objects to distinct points on the retina. How this works in detail is explored in *Dioptrice*.

Kepler's conceptual toolkit

In *Dioptrice*, Kepler assumes the following concepts without elaboration:

- Vision occurs because of the eye's sensitivity to light.

- Some objects are directly visible because they emit light. These objects are *luminous*. Other objects emit no light of their own and are visible only when illuminated by some other luminous object.

- Light is emitted in all directions from every point on the surface of either a luminous or illuminated object.

- Most materials are opaque to light. A few are transparent. Within a given transparent material, light moves along a straight line segment or ray.

- The path along which light moves may be altered by either reflection or refraction. Refraction occurs when light exits one transparent material and enters another.

- If light is reflected or refracted in transit, the direction along which it enters the eye may not be the same as the direction from object to eye. But we have no independent knowledge of the actual path taken by light entering our eyes. Consequently, our perception is that objects actually lie in that direction from which their light enters the eye.

Refraction

Dioptrice contains many diagrams depicting the refraction of light. Typically, light is either entering glass from air or exiting glass into air. The profile of the glass surface is indicated by a line or curve. The path along which the light travels is then indicated by two line segments, referred to as *rays*.

Figure 1.5 illustrates the refraction of light passing from air into glass. Light approaches the glass surface AC along the incident ray EB. The direction BG of the light inside the glass differs from the original direction EBF. Kepler calls the angle ρ (rho) between these two directions the *angle of refraction*. Note that angles DBE and FBH are both supplements of angle EBH, and thus equal. Equality of these angles is indicated in the figure by the two arcs with single hash marks.

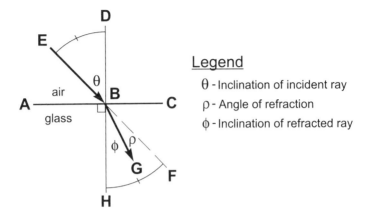

Figure 1.5 The geometry of refraction.

Line DH is the perpendicular to surface AC at point B. Kepler calls the angle θ (theta) between the incident ray EB and the perpendicular DB the *inclination* of the incident ray. Similarly, angle ϕ (phi) between BG and BH is the inclination of the refracted ray.[1]

Through any point of a surface, there is only one line that can be drawn so as to be perpendicular to the surface. The perpendicular to the surface serves as a reference direction at that point. Figure 1.5 indicates that DH is perpendicular to AC by the small square

---

[1]Kepler uses the terms *inclination* and *inclination to the surface* interchangeably. In fact, the inclination of a ray is the angle between the ray and the perpendicular to the surface, not the angle between the ray and the surface itself. The term *inclination to the surface* is thus confusing, and we avoid it where possible.

at their intersection. When a surface is curved, this symbol implies perpendicularity of the line and surface only at the indicated point.

If the incident ray EB in Figure 1.5 were not refracted, but continued undeviated as BF, then EB and BF would necessarily be on opposite sides of the perpendicular DH. The fact is, though, that the refracted ray BG is deviated toward the perpendicular. The question arises whether the deviation could ever be so large as to place the refracted ray BG back on the same side of the perpendicular as the incident ray EB.

In quantitative terms, the question is whether ρ can be larger than θ. Experimentally, it is found that ρ is always less than θ. Thus, we can rely on the fact that a refracted ray is always on the opposite side of the perpendicular from the incident ray. This simple rule proves to be very useful.

Finally, a note about terminology: Kepler calls BG the *refraction* of the incident ray EB. He also says that EB is *refracted* at B. Such language is convenient, but it is important to remember that rays have no physical existence. Strictly speaking, light is refracted but rays are not. Rays are merely line segments in a diagram representing a portion of a light path.

When Kepler says that a ray is refracted, what he means is that there is a change of direction along the light path. A new ray is needed at this point to represent the new direction. The ray that starts at the

bend in the light path is the refraction of the ray that
ends there.

Chapter 2

Rules of Refraction

Kepler realized that to understand the function of the telescope, it would be necessary to understand how lenses refract light. The occurrence of refraction was well established by Kepler's time. The essential observation was that when light passes from one transparent material to another, the direction of the light may change.

Understanding refraction meant being able to predict the direction of the refracted light. It was clear that the direction of the incident light affected the direction of the refracted light. But what was the exact relation? Once Kepler identified the inclination of the incident ray and the angle of refraction as the significant geometric quantities in refraction (Figure 1.5), he was well on his way to an answer.

Kepler began his investigation by examining the refraction of light as it enters glass from air. Having specified these transparent materials, he assumed that any variation of the angle of refraction would depend solely on the inclination of the incident ray. Thus, Kepler designed an apparatus to measure inclinations of incident rays and angles of refraction. These measurements led him to a simple arithmetic rule relating the angle of refraction to the inclination.

In order to understand the use of Kepler's apparatus, you need to know what the tangent of an angle is. You also need to be comfortable with some basic concepts of plane geometry. So if you haven't been using your high school mathematics, this would be a good time to review the first two appendices.

Apparatus for measuring refraction

Kepler's apparatus appears in Figure 2.1. It consists of a glass cube ABC positioned in a wooden holder XYZ. The assembly is oriented so that the common edge AB of cube and holder is perpendicular to the rays of the Sun. Several representative rays K, L, M, and N are shown in the figure. Kepler assumes that rays from the Sun may be regarded as parallel. On this assumption, all rays that strike the top of the glass cube do so at the same inclination. Therefore, they are all refracted by the same angle.

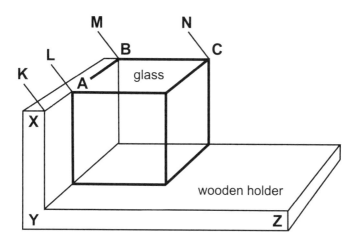

Figure 2.1 Kepler's apparatus for measuring inclination and refraction.

The sense in which rays from the Sun may be regarded as parallel merits a short discussion. Unlike other stars, the Sun does not appear to us as a point source of light. Earth is sufficiently close to the Sun that we see the Sun as a disk having an appreciable width. What matters in the present discussion is the angle between straight lines that proceed from opposite edges of the Sun to any point on earth. This angle is approximately one-half of a degree. Rays from the Sun may be regarded as parallel only when a half degree is smaller than the precision one intends to work at.

Figure 2.2 depicts the shadow that the wall XY casts on the bottom of the holder. (To simplify the drawing, only the inner surface of the holder is shown now.) The part $XADY$ of the wall not in contact with the glass cube casts shadow $QRDY$. Part DAB that is in contact with the cube also casts a shadow. But the rays passing over this portion of the wall encounter the top of the cube and are refracted downward. This results in a shorter shadow under the cube indicated by DOP.

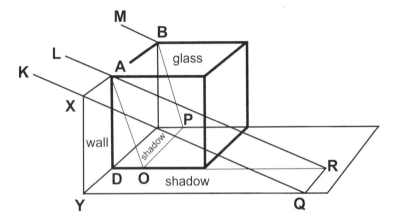

Figure 2.2 Shadows cast by the wall of the holder.

Ray LA ends at the corner A of the glass cube. If we regard LA as striking the glass just inside the corner, then it is refracted as AO. On the other hand, if we regard LA as just missing the glass, it continues unrefracted as AR. Figure 2.3 shows that the angle ρ between AR and AO is the angle of refraction.

The perpendicular to the glass surface at A is AE. Thus, the inclination of the incident ray LA is the angle θ between LA and AE. Additionally, angles LAE and

DAR are both supplements of the acute angle LAD, so they are equal. Accordingly, DAR is also labeled θ, as is LAE.

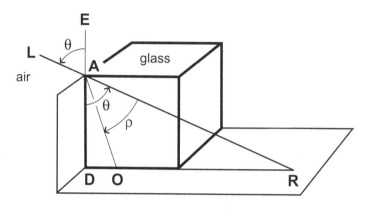

Figure 2.3 Inclination θ and angle of refraction ρ.

The apparatus is not constructed to measure θ and ρ directly. Rather, the angles are determined indirectly from the lengths of the shadows. The procedure is as follows: To find θ, the length DR of the shadow outside the cube is measured. This is facilitated by a scale inscribed on the holder. The height DA of the cube must also be measured. Then the tangent of θ is DR/DA and θ is determined from the tangent. Similarly, angle DAO is obtained from the length DO of the shadow under the cube. Finally, ρ is obtained as θ minus angle DAO.

Over the course of a day, the inclination of the incident rays varies with the height of the Sun in the sky. Thus, only a single sunny day would be required for Kepler to measure angles of refraction over a large range of inclinations. Kepler was unable to find a simple mathematical relation between inclination and refraction that was valid at all inclinations. But for inclinations ranging from $0°$ up to about $30°$, he observed that the angle of refraction was well approximated as one-third of the inclination. This simple rule is the foundation upon which all of *Dioptrice* was built.

Reversibility of ray diagrams

Kepler accepted as fact that when one draws a diagram depicting the refraction of rays at a surface, the diagram is a valid representation of light moving in either direction. Figure 2.4 depicts light moving downward through the air toward a glass surface at an inclination of θ. Refracted at the surface, the light passes into the glass where the inclination of the refracted ray is ϕ.

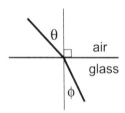

Figure 2.4 Ray diagrams are reversible.

Alternatively, the same diagram represents light moving upward inside the glass toward the surface at an inclination of ϕ. It emerges into the air at an inclination of θ. Kepler frequently exploits this reversibility of ray diagrams in *Dioptrice*.

The two rules of refraction

Given the reversibility of ray diagrams, Kepler's rule for refraction of light entering glass from air implies a second rule for refraction of light emerging from glass into air. In Figure 2.5 AB is a ray incident on the glass surface from the air. The inclination θ of the ray has been divided into three equal sectors. The divisions are continued through B into the glass to mark three corresponding sectors there.

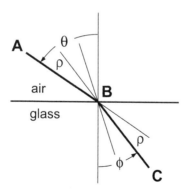

Figure 2.5 Rules of refraction.

When light enters glass from air, it is deviated toward the perpendicular by one-third the inclination of the incident ray. Thus, in Figure 2.5, the refracted ray BC has been deviated toward the perpendicular by one sector relative to the incident ray AB. Since this is the angle of refraction, it is labeled ρ.

When the diagram is interpreted in reverse, light inside the glass is directed against the surface along CB. The refracted ray in air must be BA, since ray diagrams are reversible. Thus, the light is deviated away from the perpendicular by one sector, again labeled ρ. In this case, the inclination ϕ of the incident ray CB is two sectors, so the angle of refraction is half the inclination of the incident ray.

We now see that Kepler's approximate rule of refraction is really two rules:

- Going from air to glass, the angle of refraction is one-third the inclination of the incident ray.

- Going from glass to air, the angle of refraction is one-half the inclination of the incident ray.

The rule of crossing rays

Figure 2.6 shows two rays AB and EB striking surface ST at a common point B. BA' is a continuation of AB, and is the path that light moving along AB would take if refraction did not occur. Similarly, BE' is the continuation of EB. But refraction does occur, and BC and BF are the refractions of AB and EB, respectively.

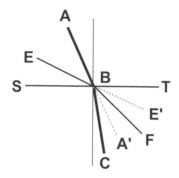

Figure 2.6 Crossing rays.

Kepler asserts that just as rays AB and EB would cross if they were not refracted, they must also cross when they are refracted. This assertion is confusing, given that rays AB and EB both terminate at B. But what Kepler means is that the complete light paths, consisting of incident and refracted rays, cross. Thus, in Figure 2.6, paths ABC and EBF cross at B.

An alternative formulation of this principle is the following: If the inclination of EB is greater than that of AB, then the refraction of EB has a greater inclination than the refraction of AB. This simple rule proves to be very useful in Kepler's hands.

Divergence

Another recurring idea in *Dioptrice* entails a collection of rays diverging from a point source and passing through an aperture. The pupil of the eye is one example of such an aperture. A lens is another. Since light is emitted in all directions from a point source, most of the rays will miss the aperture. Those rays that do enter the aperture form a solid cone with its apex at the point source and its base coincident with the aperture.

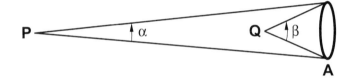

Figure 2.7 Two cones based on the same aperture.

Figure 2.7 depicts rays from two point sources, P and Q, entering the same aperture A. The ray cones originating from P and Q consist of the bounding rays, which appear in the figure, as well as all those rays in between, which do not appear. Observe that the apical angle β of the cone from Q is larger than the apical angle α of the cone from P. In general, the closer a point source is to a given aperture, the larger the apical angle of the cone of rays passing through the aperture.

Intuitively, the larger the apical angle of a ray cone, the more divergent the rays in the cone. Now, the divergence of two rays is naturally defined as the angle between them. But when we try to apply this definition of divergence to a cone of rays, there is the problem that the cone consists of a multiplicity of rays and there are different angles between different rays in the cone.

So what angle do we use? Should we take an average? Kepler uses the apical angle itself as the measure of a cone's divergence. This represents a worst-case approach. The angle between any two rays within the cone can never exceed the apical angle of the cone.

Finally, note that the smaller the apical angle of a ray cone, the more nearly parallel the rays are. Kepler often assumes that a point source is *remote*. The context of this assumption is always one of rays entering a given aperture from a point source. We have seen that the farther the point source is from the aperture, the smaller the apical angle is. Thus, assuming a point source to be remote guarantees that the rays entering the aperture are nearly parallel.

Chapter 3

The convex lens

In this chapter we follow Kepler as he undertakes a detailed analysis of the convex lens based on the rules of refraction presented in the previous chapter. Difficulties arise in the application of these rules to lenses because of their curved surfaces. Kepler's approach was to make simplifications and approximations. In this way he was able to grasp the essential features of refraction by a convex lens. Later generations would make more exact calculations and fill in the details, but the general scheme remained Kepler's.

The central concept to emerge from Kepler's analysis of the convex lens is that of a *convergence point*. Groups of rays emerging from a convex lens often converge to a good approximation on a single common point. This chapter develops the concept of the convergence point in considerable detail.

Next, the projection of images by a convex lens is explored. This is important for understanding human vision, since the lens of the eye projects an image on the retina. The chapter concludes with an account of nearsightedness and farsightedness. These variations of human eyesight are relevant to Kepler's approach to telescope design, as will be discussed in Chapter 5.

Before beginning, we need to define some terms. In *Dioptrice*, Kepler almost always assumes the faces of a lens to be either spherical or planar. When we say that the surface of a lens is spherical, it is generally the case that it comprises only a small portion of a sphere. Where the entire sphere is not present, the point corresponding to the center of the sphere is usually called the *center of curvature*, and the length corresponding to the radius of the sphere is called the *radius of curvature*. Every point on a spherical surface is the same distance from the center of curvature. That distance is the radius of curvature.

There are two ways to characterize a spherical surface. One is to state the radius of curvature. This calls attention to the size of the sphere. Alternatively, we may describe the *curvature* of the surface. The curvature is large when the radius of curvature is small. Conversely, the curvature is small when the radius of curvature is large.

By definition, a convex lens is thicker at its center than at the rim, whereas a concave lens is thinner at the center than at its rim. Galileo used one convex lens and one concave lens in his telescope. But Kepler devoted more attention to the convex lens than to the concave, and came to devise a novel telescope design using only convex lenses.

Next, an explanation is necessary regarding the figures in this book. In many situations, realistic drawings would not be useful. They would consist mostly of blank space and many important details would have to be crowded into a very small area. To avoid this, it is often necessary to draw figures in a way that does not accurately represent distances. Angles may also be distorted. For instance, it may be impractical to draw a line that is supposed to be perpendicular to a curve so that the angle is actually $90°$. Thus, the figures are not scale drawings but rather caricatures of reality, designed to convey an idea. To interpret the figures correctly, the reader will need to follow the text.

Finally, the next two sections are the most mathematically demanding in this book. It is certainly possible to read the conclusions at the end of each of these sections and then proceed without understanding how they were obtained. But if you choose to study these sections and find them difficult, you will find help in the first two appendices.

The symmetric convex lens

Kepler knew empirically that convex lenses make parallel rays converge to a point. Figure 3.1 depicts a convex lens with a collection of rays that are parallel to the *axis* of the lens. The rays strike one surface of the lens and are refracted there. The refracted rays inside the lens continue to the opposite surface where another refraction takes place. The refracted rays formed at the second surface then converge and pass through a single common point called the *convergence point of the lens*. The concept that such a point exists was a guiding principle in Kepler's study of the convex lens.

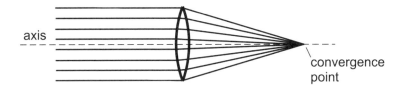

Figure 3.1 Convergence point of a convex lens.

Kepler assumed that the position of the convergence point is determined by the shape of the lens. His goal was to state precisely where the convergence point of a lens is, based on the shape of the lens. Although he was not able to solve this problem in its full generality, Kepler did obtain a solution for the case of a symmetric convex lens with spherical surfaces. This solution, in turn, provided Kepler with the guidance he needed in thinking about more complex problems.

Figure 3.2 illustrates refraction by a symmetric convex lens when the rays striking the lens are initially parallel to its axis. The surfaces of the lens are assumed to be spherical. The centers of curvature lie on the axis, which is an imaginary line passing through the center of the lens and perpendicular to both surfaces. A line drawn from a center of curvature to any point of the corresponding surface is a perpendicular to the surface at that point. Thus, in the figure, CFD is perpendicular to surface AB at point F.

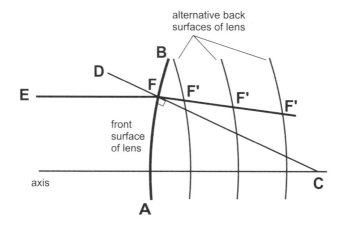

Figure 3.2 Varying the thickness of a convex lens.

When discussing the movement of light through a lens, it will be convenient to refer to the first surface encountered by the light as the *front surface*. Light entering the front surface will generally pass through the *back surface* as well. Thus, in Figure 3.2, AB will be the front surface. The figure also shows profiles of several alternative back surfaces, corresponding to different possible thicknesses of the lens.

Ray EF is representative of a set of rays parallel to the axis of the lens. These strike the front surface of the lens from the left. EF is refracted at the front surface, and the refracted ray strikes the back surface at a point F'. The alternative locations for F' indicated in the figure depend on where the back surface of the lens is located.

Relative to the incident ray EF, the refracted ray FF' inside the lens is deviated toward the perpendicular FC. So, unlike EF, FF' is inclined to the axis, and F' will be closer to the axis than F is. But how much closer depends on how thick the lens is. If the lens is very thin, then F' will be only slightly closer to the axis than F is. To simplify his calculations, Kepler assumes that F' is negligibly closer to the axis than F. In other words, he assumes the lens to be thin.

Figure 3.3 depicts refraction by a thin symmetric lens. Again, ray EF represents any of a collection of rays parallel to the axis striking the lens from the left. The centers of curvature of the front and back surfaces are points C and A, respectively, and since the lens is symmetric, A is the same distance from the back surface as C is from the front surface.

Line CFD is the perpendicular to the front surface at the point F where ray EF strikes the lens. Line $AF'B$ is the perpendicular to the back surface at the point F' where the refraction FF' strikes the back surface. If the lens is sufficiently thin, we can neglect the fact that F' is closer to the axis than F, and regard $EFF'G$ as a straight line continuing EF through the lens.

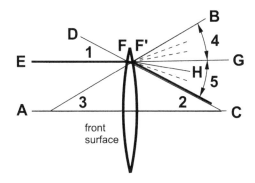

Figure 3.3 Parallel rays refracted by a symmetric lens.

Kepler proves that all of the rays striking the lens parallel to the axis will be refracted through point C, one radius of curvature from the front surface. The first step in the demonstration is to show that angles $GF'B$ and GFC are each equal to the inclination DFE of the incident ray EF. This is slightly tedious, but not difficult.

DFC is a transversal cutting parallels EG and AC, so angle DCA equals DFE. Then, since the lens is symmetric, angle BAC equals DCA. Similarly, $BF'A$ is a transversal cutting parallels EG and AC, and $GF'B$ equals BAC. Finally, GFC and DFE are both supplements of DFG and are thus equal. To summarize, those angles in the figure labeled 1 through 4 are equal and angles 1 and 5 are also equal, so angles 4 and 5 are equal, which is what we wanted to show.

Now, DFE is the inclination of the incident ray EF to the front surface. Since the light is entering glass from air, the angle of refraction will be one-third of DFE. Relative to the incident ray EF, the refraction FF', will be inclined toward the perpendicular DFC by one-third of DFE.

It is difficult to appreciate the direction of FF' in the figure, so we extend FF' to H beyond the back surface and use $F'H$ to represent the direction of FF'. Recalling that angles GFC and DFE are equal, we divide GFC into three equal parts. Then FH lies at the boundary between the first and second parts, as indicated in the figure.

At this stage of the demonstration, Kepler assumes the thickness of the lens to be negligible and treats F and F' as the same point. But while the length of FF' is negligible, its previously determined direction $F'H$ remains significant and unchanged. Finally, due to the identification of F and F', we now regard angles GFC, $GF'C$, and $GF'B$ as equal to one another and equal to the inclination DEF.

We now turn our attention to the refraction of FF' at the back surface of the lens. The perpendicular to the back surface at F' is $F'B$, so the inclination of FF' to the back surface is the angle $HF'B$. We have already divided $GF'C$ into three equal parts. We now divide $GF'B$ into three equal parts, equal to those of $GF'C$. The inclination $HF'B$ equals four of these parts.

What trajectory does the refraction of FF' take? Going from glass to air, the angle of refraction is one-half the inclination of the incident ray. Half of $HF'B$ is two subdivisions. These two subdivisions add to the four of $HF'B$ because, going from glass to air, the refraction is away from the perpendicular. So the refraction of FF' at the back surface lies along a line six subdivisions from the perpendicular $F'B$, which is to say, along $F'C$. Consequently, the final refraction intersects the axis at C.

What has been said about the refraction of EF applies without change to any of the parallel rays striking the lens. Thus, C is the convergence point of refractions produced from all these rays.

Conclusion: A symmetric convex lens refracts rays parallel to its axis through a convergence point one radius of curvature from the front surface.

Refraction of parallel rays by a single spherical surface

The previous section discussed the refraction of parallel rays striking a symmetric lens from the outside. Here, rays parallel to the axis are assumed to be somehow already present inside a convex lens. It will

be shown that they are refracted by a spherical surface
to a point two radii of curvature beyond the surface.

In Figure 3.4 AB is the spherical surface of a
convex lens and C is the center of curvature. The axis
of the lens is CBF. Line DE is parallel to the axis and
DA is any one of the parallel rays inside the lens. The
ray is refracted at A so as to pass through F. We want to
confirm that all rays parallel to DA will pass through
the same point F, and we want to know the distance
BF from the lens to point F.

First, we establish various relations between the
angles in the figure. The perpendicular to the surface
at A is CAG, and the inclination of ray DA to this
perpendicular is the angle DAC, also labeled θ. Note
that the closer DA is to the axis, the smaller θ will be.

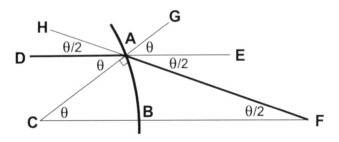

Figure 3.4 Parallel rays emerging from a lens.

Angles DAC and GAE are equal because they are
both supplements of angle DAG. GAC is a transversal
cutting parallels DE and CF, so angle ACB is also
θ. The refraction of DA is AF, and because light is

emerging from glass into air, the angle of refraction EAF is $\theta/2$. HAD and EAF are equal because they are both supplements of HAE. Finally, transversal HF makes AFC also equal to $\theta/2$.

Next, we find the length of CF by applying the rule of sines to triangle CFA. For this, we need the sine of angle CAF. Now, the sine of an angle equals the sine of its supplement. The supplement of CAF is CAH, which is $3\theta/2$. Then the rule of sines gives

$$sin(CAF)/CF = sin(CFA)/CA,$$
$$sin(3\theta/2)/CF = sin(\theta/2)/CA.$$

If the rays inside the lens are close enough to the axis, angles $3\theta/2$ and $\theta/2$ will be small and the sines and angles will be proportional. Then,

$$(3\theta/2)/CF = (\theta/2)/CA,$$
$$3/CF = 1/CA,$$
$$CF = 3CA.$$

Examination of the figure shows that $BF = CF - CB$. Then, because CA and CB are equal,

$$BF = 3CA - CA$$
$$= 2CA.$$

This result in no way depends on θ, so despite the fact that θ is different for each ray inside the lens, the refractions of all of them pass through F.

Conclusion: Any ray inside a convex lens that is close to the axis and parallel to it is refracted through a point two radii of curvature from the surface of the lens.

Refraction of rays from a point source

The result of the previous section is readily extended to the case of an asymmetric lens. To do this, we again use Kepler's artifice of assuming that rays parallel to the axis are somehow present inside the lens. Figure 3.5 illustrates the case in which the radius of curvature of the surface on the left is larger than that on the right. The parallel rays inside the lens are refracted at both surfaces so as to converge at points two radii of curvature from the surface.

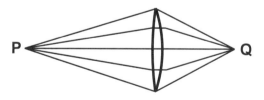

Figure 3.5 Parallel rays emerging from an asymmetric lens.

The importance of this example becomes clear if, instead of tracing the rays in both directions from inside the lens, the rays are traced from one convergence point to the other. Now the interpretation becomes that rays from a strategically located point source P on one side of the lens are refracted to a convergence point Q on the other side of the lens. This is something new. Until now we have spoken only of the convergence point of the lens, that point to which a convex lens refracts rays parallel to its axis. Here we have a convergence point associated with a point source.

The question arises, would rays from a point source on the axis that is not two radii of curvature from the lens also have a convergence point? What if the point source is not on the axis? The answer in both cases is yes, approximately, provided that the relevant angles are not too large. This affirmative answer is what makes the Al-Kindian conception of a visual scene as an assemblage of point sources so powerful. To every point source on the surface of a visible object, there corresponds a unique convergence point on the opposite side of the lens.

On-axis source and convergence point

In this section, we address those cases where the point source is on the axis of a convex lens. The following section will treat the case of the off-axis point. To make the discussion as simple as possible, we assume a symmetric lens. The results obtained can easily be extended to asymmetric lenses.

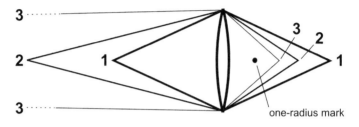

Figure 3.6 Source points and convergence points.

In Figure 3.6 point 1 on the left represents a point source that is two radii of curvature from a symmetric convex lens. Light from this point converges to a point, again labeled 1, two radii of curvature to the right of the lens. Now, let the point source on the left become progressively farther from the lens, as illustrated by points 2 and 3, the latter being too far from the lens to appear in the figure. What is found experimentally is that, as the point source on the left becomes more distant from the lens, the corresponding convergence point on the right becomes progressively closer to the lens.

Although this general trend is simple, the detailed description of what happens is a bit subtle. As long as the point source recedes only a small distance from position 1 on the left, the convergence point on the right approaches the lens by a comparable distance. But when the point source is already some appreciable distance to the left of position 1, it takes a large displacement of the source to produce a small displacement of the convergence point. In fact, the disparity between the displacement of the source and that of the convergence point grows so as to guarantee that, no matter how far the source goes from the lens, the convergence point never reaches the one-radius mark.

Subtleties aside, here is what you should remember: If a point source is very distant from a lens and on the axis, its convergence point will practically coincide with the convergence point of the lens. But in general, the convergence point of a source is farther from the lens than the convergence point of the lens.

Now, invoking the reversibility of ray diagrams, let the points in Figure 3.6 to the right of the lens be interpreted as sources. When the source is two radii of curvature from the lens, at position 1, the corresponding convergence point 1 is two radii of curvature to the left of the lens. When the point source is closer to the lens, at position 2, the corresponding convergence point 2 is farther to the left. With the source close to the one-radius mark, such as at position 3 on the right, the corresponding convergence point is very far to the left. Finally, if the point source were to be placed precisely at the one-radius mark, the rays emerging from the opposite surface would be parallel, and no convergence point would exist.

Nothing prevents the placement of a point source even closer to the lens than the one-radius mark. Figure 3.7 illustrates the case of a point source P located between the lens and the one-radius mark. Despite the fact that the divergent rays from the point source become less divergent after refraction at the first surface, and less divergent still when refracted at the second surface, the rays emerging from the lens remain divergent. Just as for a point source at the one-radius mark, there is no convergence point.

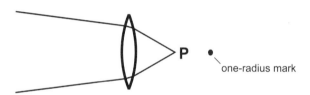

Figure 3.7 Distance between source and lens less than one radius of curvature.

Incidentally, Figure 3.7 provides a nice summary of how convex lenses function. Tracing light paths from left to right, we have convergent rays that are made more convergent by the convex lens. Tracing light paths from right to left, the lens makes divergent rays less divergent.

Off-axis source and convergence point

Point sources located on the axis of the lens have corresponding convergence points that are also on the axis. The distinctive feature of a point source not lying on the axis is that the corresponding convergence point is also off-axis. Additionally, the source and convergence point are always on opposite sides of the axis. Kepler assumes this important fact without discussion. We will show how this results from the fact that incident and refracted rays are always on opposite sides of the perpendicular.

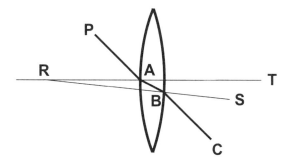

Figure 3.8 Refraction of a ray from an off-axis source.

In Figure 3.8 a ray is drawn from an off-axis point source P to the point A, where the front surface of the lens intersects its axis RT. The perpendicular to the surface at A is the axis itself. The incident ray PA and the refracted ray AB must be on opposite sides of the perpendicular. In this case, that is the same as being on opposite sides of the axis.

What has to be shown is that the refraction that takes place at the back of the lens will not produce a refracted ray that goes back across the axis. If R is the center of curvature of the back surface, the perpendicular to the surface at B is RBS. Since B is below the axis in the figure, RBS slopes downward from point R on the axis. The refracted ray BC must be on the opposite side of RBS from AB, and this guarantees that it lies below the axis in its entirety.

Now, on the assumption that a convergence point Q exists, by definition, all rays from P that strike the lens are refracted so as to pass through Q. Stated differently, the convergence point lies on every refracted ray. But this includes ray BC, which has been shown to lie entirely below the axis. Therefore, the convergence point Q is also below the axis, as shown in Figure 3.9. We conclude that an off-axis point source and its corresponding convergence point are located on opposite sides of the axis.

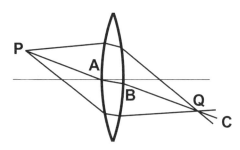

Figure 3.9 Off-axis source and convergence points.

In other respects, off-axis point sources behave like on-axis sources. Just as with on-axis points, the farther an off-axis source is from the lens, the closer its convergence point is to the lens. Also, the convergence point of an off-axis source is never closer to the lens than the convergence point of the lens.

Projection of an image by a convex lens

We now consider something more complex than a single point source and its corresponding convergence point. This section discusses the use of a convex lens to project the image of an object onto a surface. In the next chapter, we will explore the appearance of objects viewed through a convex lens. In both cases, the analysis depends on the conception of a visible object as a multiplicity of source points. To each point source there is a corresponding convergence point. These convergence points are located in a limited region of space on the opposite side of the lens from the object.

Try this. In a darkened room, use a convex lens to project the image of a luminous object, such as a lamp, on a piece of white paper. A cheap magnifying glass will work well. Position the lens across the room from the object so that the object is remote. Then, by trial and error, find the location behind the lens where the image on the paper is most clear.

The position at which the best image is seen on the paper is the location of the convergence points. At each convergence point, light from the corresponding object point coalesces and illuminates the paper held there. The arrangement of the convergence points reproduces the arrangement of object points, so an image of the object appears on the paper.

But as you move the paper from this location, the image becomes increasingly blurred. The reason is that light from each object point now illuminates an extended area of the paper rather than just one point. At the same time, any point of the paper is illuminated by light from a multiplicity of object points.

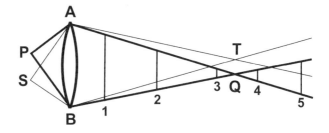

Figure 3.10 Image formation by a convex lens.

We illustrate this in detail with the help of Figure 3.10. In the figure, rays from object points P and S diverge against a convex lens. The refracted rays converge to points Q and T, respectively, after which they diverge once again.[1]

---

[1]Note that, because of space constraints, points P and S are

Now, suppose a piece of paper were to intercept the converging ray cone AQB at position 1. The light from P would be dispersed over a large circular area on the paper, the profile of which is represented by the vertical line segment. It is evident that the majority of this area would also be illuminated by ray cone ATB from object point S.

Progressing toward the convergence points, the areas illuminated by each source point become progressively smaller, as indicated for cone AQB by the shorter lengths of line segments 2 and 3. Note that the overlap between cones AQB and ATB is less at position 2 than at position 1. At position 3 these cones no longer overlap at all. But the fact is that ray cones from sources that are closer to one another than P and S still would overlap even here.

At Q and T, the ray cones pass through their respective convergence points. Here, there is no overlap of cones, no matter how close the object points are to one another. We see a distinct image on a surface placed here because each point on the surface is illuminated by a single object point. Finally, progressing beyond Q and T, the ray cones become divergent. Light from each point source is again dispersed over an area that increases with distance from the lens, as indicated by the increasing lengths of line segments 4

---

drawn close to the lens in the figure. But their distance from the lens must actually be greater than the distance of the lens to its own convergence point. Otherwise, light from these points would still be divergent after passing through the lens.

and 5.

In summary, an ideal image requires a one-to-one correspondence between object points and image points. There is a ray cone at the back of the convex lens corresponding to each object point. If a surface is placed so as to intercept the ray cones at their apices, then the one-to-one correspondence is established. But with the surface at any other location, each object point illuminates many points on the surface and each point on the surface is illuminated by many object points. The result is a blurred image.

Images projected by a convex lens are inverted. The inverted images produced in a camera obscura were explained by invoking the fact that light moves along straight lines. In contrast, the inversion of the image projected by a convex lens is explained by the fact that convergence points are always on the opposite side of the axis from the corresponding object points. Since an image is perceived only in the region of the convergence points and the convergence points are inverted with respect to the object points, the image projected by a convex lens is always inverted.

Deficiencies of vision

Having studied the operation of the convex lens, Kepler uses the knowledge gained to explain near-sightedness and farsightedness. The names given to these conditions can be confusing. We define the conditions as follows: A person who is nearsighted sees nearby objects clearly, but not distant objects. A farsighted person sees distant objects clearly, but not objects that are nearby.

Kepler explains that distant objects appear blurred to a nearsighted person because the retina is far from the convergence points of distant objects. Is the problem that the retina is closer to the lens than the convergence points, or is it farther away? The answer lies in the fact that this person sees nearby objects clearly.

We know that convergence points of nearby objects are farther from the lens than those of distant objects. So it must be that the retina is too far from the lens to see distant objects clearly. As illustrated in Figure 3.11, rays from a nearby object point P converge at D, which lies on the retina. But rays such as QA and QB from a distant object point Q converge between the lens and retina at C.

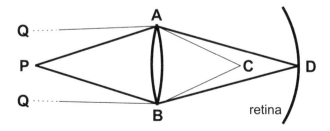

Figure 3.11 The nearsighted eye.

This explanation suggests, however, that the eye can see objects clearly only at one specific distance. In fact, the eye has a means of seeing clearly over a range of distances. This is called *accommodation*. Kepler guessed that, just as the pupil of the eye is able to constrict and dilate, there must be a mechanism for adjusting the distance between the lens and retina. If so, a more precise description of nearsightedness would be that the nearsighted eye is unable to bring the retina sufficiently close to the lens for distant objects to be seen clearly.

Kepler indulged in some speculation as to how this situation could arise. He leaned to the opinion that nearsightedness is an acquired condition that results from constantly viewing objects at close quarters. Similarly, farsightedness would result from constantly looking far away.

We now know that the human eye adjusts the shape of the lens, rather than the separation between lens and retina, in order to see clearly over a range of distances. But Kepler was not far from the mark. He correctly understood why such a mechanism is needed and what it accomplishes. In fact, the eye of the octopus does operate precisely as Kepler suggested.

The current understanding is that the normal eye increases the radius of curvature of the lens in order to shift the convergence points of distant objects back to the retina. That is to say, it reduces the curvature of the lens. Similarly, by increasing the curvature of the lens and reducing its radius of curvature, the convergence points of nearby objects are shifted forward to the retina.

With this understanding, the explanation of either nearsightedness or farsightedness is that there is a mismatch between the achievable curvatures of the lens and the fixed distance between the lens and retina. In nearsightedness, the radius of curvature cannot be made large enough to shift the convergence points of distant objects back to the retina. In farsightedness, the radius of curvature cannot be made small enough to shift the convergence points of nearby objects forward to the retina.

Incidentally, there is another way in which our current understanding of the eye differs from Kepler's. He believed that the lens of the eye was the principal refractive element in the eye. We now know that the cornea of the eye actually refracts light more powerfully than the lens. The true role of the lens is to fine-tune the amount of refraction. For simplicity, and to remain within Kepler's conceptual framework, we will ignore the cornea and speak as if all the refraction in the eye occurs at the lens.

Chapter 4

Looking through a convex lens

Besides projecting an image by means of a convex lens, it is possible to view objects directly through the lens, as one does with a magnifying glass. Anyone who handles a magnifying glass for more than a few moments will observe some puzzling phenomena. Depending on the position of the eye and the object in relation to the lens, the object may appear larger or smaller than it appears without the lens. It may appear in its true orientation or upside-down and backward. It may be clearly seen, or it may be blurred. The object may even fail to appear through the lens altogether, despite the fact that the lens is held directly between the object and the eye. This chapter will explain all of these phenomena.

Blurring

There is a surprising difference between projection of images by a convex lens and observation of objects through the lens. Images projected by a convex lens onto a surface are clear only when the surface is in the vicinity of the convergence points. In contrast, if the eye observes an object from this region, the object appears badly blurred. But if the eye moves closer to the lens than the convergence points, or farther away, the object is seen more clearly. Why?

In Chapter 3 we discussed why a projected image may be blurred. Away from the convergence points, the cone of rays from a given object point illuminates an extended region of the surface (Figure 3.10). At the same time, each point of the surface is illuminated by overlapping ray cones from many object points.

Ultimately, blurring occurs for the same reason when an object is directly viewed through a convex lens. Blurring occurs when ray cones from different object points overlap on the retina. But this criterion is more difficult to apply here because rays must be traced through both the external lens and the lens of the eye before one can know where the rays will strike the retina. Kepler avoided this complication by using the degree of convergence or divergence of rays entering the pupil to predict the amount of blurring the eye will perceive.

When the eye views a distant object, rays from

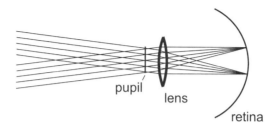

Figure 4.1 Two bundles of parallel rays overlapping at the pupil.

each object point are, to a good approximation, parallel. Figure 4.1 depicts bundles of parallel rays entering the pupil from two different directions. This could represent, for example, light entering the eye from two stars. Then, although light from the two stars is completely overlapping at the pupil, the lens sorts out the rays, dispatching them to distinct points of the retina so that the stars are seen as distinct points of light.

Figure 4.2 depicts an eye viewing two nearby point sources P and Q with convergence points S and T, respectively. We know that the farther from a convex lens a point source is, the closer to the lens is its convergence point. In this case, P is sufficiently distant from the lens that convergence point S is close to the retina. It is brought precisely to the retina by appropriate adjustments to the curvature of the eye's lens. Consequently, P is seen clearly.

But Q is too close to the eye to be seen clearly.

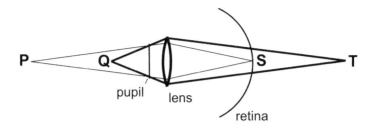

Figure 4.2 Ray cones from nearby point sources.

Even when the eye shortens the radius of curvature of its lens maximally, convergence point T remains behind the retina. Consequently, Q illuminates an extended area of the retina and Q appears as a blurred disk rather than as a point.

At the end of Chapter 2, we defined the divergence of a ray cone as the apical angle of the cone. We also noted that the closer a point source is to an aperture, the larger the divergence of the cone of rays entering the aperture. The cone of rays from Q entering the pupil has a greater divergence than the cone from P. So instead of saying that Q is too close for the eye to see it clearly, perhaps it could be argued that the rays are too divergent for the eye to see Q clearly.

This was the approach that Kepler took. Blurring occurs when the divergence of rays from an object point exceeds the limit of what the eye can handle. The more the divergence exceeds this limit, the greater the blurring.

How does this work out in detail? The key is to think in terms of curvature rather than radius of curvature. (Recall that the curvature increases as the radius of curvature decreases.) A convex lens is capable of producing convergent rays from divergent rays. But given a lens of fixed curvature, the more divergent the rays entering the lens, the less convergent the rays exiting the lens.

Thus, to produce rays that converge within a specified distance of the lens, the curvature of the lens must increase if the divergence of the rays increases. But there is a limit to how much the eye can increase the curvature of its lens. It follows that there is a corresponding limit to how divergent the rays from a point source may be if they are to be made to converge at a point on the retina.

Kepler points out that the natural state of affairs is for the eye to receive divergent light from the environment. This is based on the Al-Kindian concept of visible objects as aggregations of point sources. The eye simultaneously receives light diverging from a multitude of points. Of course, the eye also encounters rays that for all practical purposes are parallel when it views distant objects. But in the absence of a convex lens, rays entering the eye are never convergent.

Figure 4.3 depicts the interaction of the eye with convergent rays from a convex lens. Initially divergent rays from a point source are rendered convergent by the lens, and these in turn are made even more

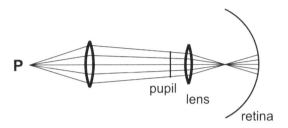

Figure 4.3 Convergent rays entering the pupil.

convergent by the lens of the eye. The rays pass through their convergence point and diverge again before striking the retina. The eye perceives the point source as a blurred disk, just as in the case of rays entering the pupil with a divergence too great for the eye.

Note that the degree of convergence of a cone of converging rays can also be defined as the apical angle of the cone. It is evident that the more convergent the rays entering the pupil are, the closer their convergence point will be to the lens of the eye, and the more they will have diverged again before striking the retina. Therefore, the degree of blurring increases with the degree of convergence.

With this understanding, let's examine in greater detail how the blurring of an object viewed through a convex lens depends on the distance between the lens and the eye. In Figure 4.4 a cone of rays diverges from P against lens AB. These rays are refracted by the lens so as to pass through convergence point Q. The cone

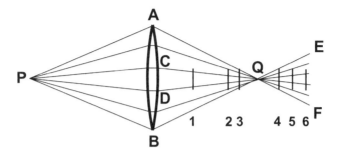

Figure 4.4 Viewing a point source through a convex lens.

of refracted rays exiting the back of the lens is AQB. After converging to Q, the rays diverge again as cone EQF. The vertical line segments labeled 1 through 6 represent the size and position of the pupil.

When the pupil is at 1, the full cone of rays AQB cannot enter the eye. Those rays that do enter constitute cone CQD, which has a smaller apical angle than AQB. Thus, the collection of rays entering the eye at this position has a smaller convergence than the complete set of rays exiting the lens.

At position 2, a greater portion of the rays from the lens enters the pupil and the eye has to contend with a greater degree of convergence. Finally, at position 3, the pupil admits the entire set of rays exiting the lens. Between position 3 and Q, there is no further increase in the convergence of rays entering the pupil because the same set of rays is admitted throughout this region.

Beyond Q, the rays are divergent and the degree of divergence is given by the apical angle of cone EQF. At all positions between Q and 4, the same set of rays enters the pupil and the divergence is constant. Thereafter, as the eye proceeds farther from the lens, the divergence of the rays entering the pupil becomes steadily smaller.

In this way, Kepler correlates the amount of blurring perceived by the eye at different distances from a convex lens with the amount of convergence or divergence at these distances. As the eye recedes from a position close to the lens, the rays entering the eye become increasingly convergent and, simultaneously, objects appear increasingly blurred. This trend continues until the eye reaches the vicinity of the convergence points where convergent rays become divergent. With further distance from the lens, the rays entering the eye become progressively less divergent and, simultaneously, the blurring decreases.

The way in which convergence and divergence vary with distance from a convex lens is the key to understanding the difference noted above between projection of images by a convex lens and observation of objects through the lens. Again, projected images are blurred in those regions where directly viewed objects appear clear. Conversely, directly viewed objects are blurred where projected images are clear.

The explanation is that formation of a clear image depends on the illumination of different points on a surface by rays from different object points. This requirement is fulfilled near the convergence points, where rays from different object points are well separated in position. But it is precisely in the vicinity of the convergence points where the convergence or divergence of rays is greatest. Under these conditions, objects viewed through the lens are badly blurred.

At locations away from the convergence points, rays from a given object point are more uniform in direction and the eye sees objects more clearly. But where the convergence or divergence is small, there is poor spatial separation of rays from different object points. Under these conditions, projected images are badly blurred.

Orientation

The appearance of objects viewed through a convex lens depends on the position of the eye in yet another way. When the eye is between the lens and the convergence points, objects appear in their true orientation. But objects appear inverted when viewed from beyond the convergence points.

To keep things simple, we have assumed the eye to be positioned on the axis of the lens. We also assume that conditions are such that the eye sees the object with reasonable clarity. This requires that the direction of rays from any object point be relatively uniform at the pupil. Referring back to Figure 4.4, it is evident that this is possible only if the pupil is small compared to the lens and either close to the lens in comparison to the convergence points, or a good deal farther from the lens than the convergence points. As usual, the diagrams that follow do not represent distances and angles accurately. The pupil is drawn rather large and neither very close to nor very far from the lens.

Suppose that the eye is between the lens and the convergence points. We will follow the light paths for a representative off-axis object point. There will be two cases to consider, depending on whether the corresponding convergence point is farther from or closer to the axis than the edge of the pupil is.

In Figure 4.5 the point source P is sufficiently close to the axis that its convergence point Q is closer to the axis than is the lower edge of the pupil. Three light paths are drawn from object point P through the convergence point Q. Paths PAQ and PBQ pass through opposite edges of the lens. Since the convergence point is closer to the axis than the edges of the lens, ray AQ slopes downward and ray BQ upward.

The rays in the convergent cone AQB vary smoothly in direction as one progresses from the down-going ray AQ to the up-going ray BQ. Therefore, there must be exactly one ray CQ in the ray cone that leaves the lens at C and maintains a constant distance

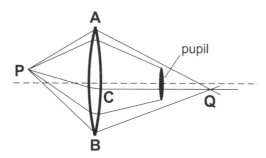

Figure 4.5 Convergence point beyond pupil, close to axis.

from the axis as it passes into the pupil. By assumption, CQ doesn't just graze the edge of the pupil, but enters the pupil more centrally. Therefore there are other rays that enter the pupil both above and below CQ. Those that enter above are down-going, and those that enter below are up-going. Consequently, object point P will be perceived as a blurred disk that straddles the axis. How disk-like or how point-like the object point appears will depend on the range of directions from which its light enters the pupil.

Figure 4.6 illustrates the second case, in which P is sufficiently distant from the axis that its convergence point is farther from the axis than the edge of the pupil. Now ray CQ doesn't enter the pupil, and all rays that do enter the pupil are down-going. Consequently, object point P appears above the axis. To summarize, when the pupil is closer to the lens than the convergence points, object points either appear to straddle the axis or to lie on that side of the axis where they actually are.

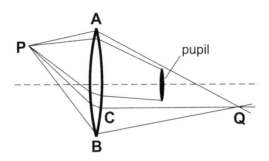

Figure 4.6 Convergence point beyond pupil, far from axis.

We repeat the same analysis for an eye that is farther from the lens than the convergence points. In Figure 4.7 object point P is close to the axis so that its convergence point Q is closer to the axis than the lower edge of the pupil is. Ray CQ enters the pupil horizontally. Rays entering the pupil from above CQ are down-going, and rays entering the pupil from below CQ are up-going. Thus, P appears disk-like and straddles the axis.

Figure 4.8 illustrates the second case in which P is sufficiently far from the axis that its convergence point Q is farther from the axis than the lower edge of the pupil is. Now ray CQ doesn't enter the pupil, and all rays that do enter are up-going. Consequently, object point P appears below the axis. Thus, when the pupil

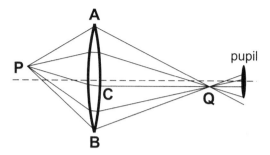

Figure 4.7 Convergence point before pupil, close to axis.

is farther from the lens than the convergence points, object points either appear to straddle the axis or to lie on the opposite side of the axis from where they actually are.

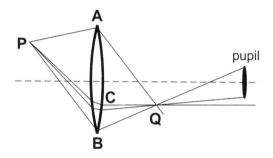

Figure 4.8 Convergence point before pupil, far from axis.

To summarize, when the eye views an object from between a convex lens and the convergence points, the object appears in its true orientation. But when the eye is farther from the lens than the convergence points, the object appears inverted.

Experience with a convex lens

In the next chapter, we will discuss how a convex lens produces magnification. In anticipation of this, and as a review of the current chapter, the reader is encouraged to obtain a convex lens and observe a distant object through it. A magnifying glass is perfectly adequate.

The convergence point of a typical magnifying glass is about 25 cm (10 inches) from the lens. Press the lens to your eye, or if you wear glasses, touch the lens lightly against one lens of your glasses, and view a distant object. The object appears in its true orientation, but somewhat blurred. Next, draw the lens farther from the eye, keeping the object in view through the lens. The object appears progressively larger and more blurred until, when the separation between lens and eye is roughly equal to 25 cm, nothing more than a blur of light is seen. Then, hold the lens at arm's length and view the same object. It now appears smaller than without the lens, inverted, and without appreciable blurring.

Chapter 5

The telescope

In this final chapter, we present Kepler's explanation of Galileo's telescope. While working out a theory of lenses, Kepler arrived at an original telescope design of his own, and this is also discussed. For Galileo and his contemporaries, the essential quality of the telescope was its ability to magnify distant objects. Thus, we begin with Kepler's explanation of magnification by a single convex lens.

Magnification

When observing an object through a convex lens, the position of the eye in relation to the convergence points has three consequences:

- Blurring is maximal in the vicinity of the convergence points and less elsewhere.

- Beyond the convergence points, objects appear inverted, whereas objects appear in their true orientation when they are between the lens and the convergence points.

- The apparent size of an object viewed through a convex lens is dependent on the position of the eye.

We discussed the first two points in the previous chapter. We now discuss the third.

When the eye is close to the lens, objects appear the same size as when viewed without the lens. But as the eye moves toward the convergence points, objects appear progressively larger. Beyond the convergence points, the opposite trend sets in. The magnification diminishes until the object again appears the same size through the lens as without the lens. With further distance still, the object appears progressively smaller.

Kepler associates the apparent size of an object with the angle that rays from opposite edges of the object form at the eye. He shows that this angle may be increased by means of a convex lens. In Figure 5.1 PQ represents an object, and the eye is at E, between the lens and its convergence point R. In the absence of the lens, rays PE and QE from the edges of the object would enter the eye along straight lines and the angle subtended by the object would be PEQ.

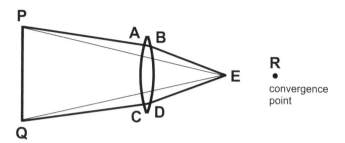

Figure 5.1 Magnification by a convex lens.

With the lens present, however, rays from P and Q cannot follow straight paths to the eye. Rather, light from P makes its way to the eye along path $PABE$, and light from Q along path $QCDE$. With the lens present, BED is the angle at the eye between rays arriving from opposite edges of the object. Since angle BED is greater than angle PEQ, the object appears larger with the lens than without the lens.

The Keplerian telescope

The blurring of objects viewed through a convex lens was explored in Chapter 4, and nearsightedness and farsightedness were considered in Chapter 3. From these discussions, it is clear that a telescope needs to provide the user's eye with parallel or slightly divergent rays for the observed object to be seen clearly. Most people's eyes can deal with parallel rays and rays that are slightly to moderately divergent. Farsighted users will be comfortable with parallel rays, but unable to cope with significant divergence. In contrast, nearsighted users cannot handle parallel rays, and require a substantial degree of divergence to see clearly.

These considerations led Kepler to a design for a telescope that differed from Galileo's. The aim was to take advantage of the magnification available near the convergence points without being confounded by the blurring that normally takes place there. Recall that blurring is due to the convergence of rays when the eye is between the lens and the convergence points. In contrast, blurring is due to excessive divergence of rays when the eye is beyond the convergence points. Kepler realized that a second convex lens placed just beyond the convergence points would serve to diminish the divergence of rays and render them suitable for viewing.

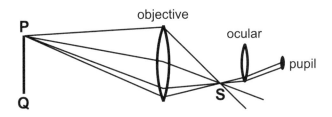

Figure 5.2 Layout of the Keplerian telescope.

The layout of Kepler's telescope is shown in Figure 5.2. An object PQ is observed through two lenses. The lens closest to the object is called the *objective* and that closest to the eye is called the *ocular*. Rays from object point P are shown diverging against the front surface of the objective lens. Their refractions pass through the convergence point S. Only those refractions from the lower portion of the objective lens strike the ocular lens and are refracted into the pupil of the eye.

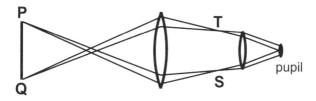

Figure 5.3 Light paths in the Keplerian telescope.

Figure 5.3 traces light paths into the pupil from points P and Q at opposite edges of the object. Rays from Q converge toward point T, while rays from P converge toward S. Kepler places the ocular lens beyond the convergence points. After passing through S and T, the rays diverge against the ocular.

Recall from Chapter 4 that when the eye is positioned beyond the convergence points, rays from an object point on one side of the axis enter the eye from the other side of the axis. In the Keplerian telescope, the ocular lens is interposed between the convergence points and the eye. Although the ocular does modify the light paths, the rays still enter the eye from the opposite side of the axis from where they originated. Consequently, objects appear inverted when viewed through the Keplerian telescope.

Convex lenses make divergent rays less divergent. This is one function of the ocular. It refracts the rays diverging from the convergence points to produce rays that are almost parallel. These enter the pupil and the eye sees without blurring. A convex lens also makes convergent rays more convergent. This is the second function of the ocular and is the basis of magnification by the telescope.

In Figure 5.3 above, two rays from each of S and T strike the ocular. Consider the one ray from each of S and T that strikes the ocular closest to its edge. These two rays converge toward the axis and are made more convergent by the ocular. They become more steeply inclined to the pupil, thereby increasing the angle that object PQ subtends when viewed through the telescope.

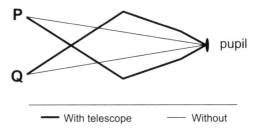

Figure 5.4 Light paths with and without the telescope.

To convey this idea more clearly, Figure 5.4 reproduces two of the light paths from Figure 5.3.

Lines are also drawn from the edges of object PQ to the center of the pupil to indicate the angle that PQ would subtend at the eye without the telescope. The telescope magnifies because rays from opposite edges of the object subtend a greater angle at the pupil with the telescope than without.

The concave lens

Kepler's treatment of concave lenses is less complete than that of convex lenses, and we will also be brief here. Concave lenses make divergent rays more divergent and convergent rays less convergent. We will demonstrate the plausibility of this claim by considering a particular configuration of rays inside a concave lens.

Figure 5.5 depicts a concave lens, inside which there are two rays that are parallel to one another, though not necessarily to the axis of the lens. Ray CE has been chosen to be perpendicular to the left-hand surface. Thus, light moving to the left along this ray would emerge from the glass without deviation as CA. A point F has been found on the right-hand surface where the perpendicular to the surface is parallel to CE. Ray DF, coinciding with this perpendicular, emerges to the right as FH without deviation.

Additional perpendiculars to the surfaces are drawn at D and E. EG is the refraction of CE at E.

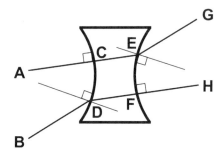

Figure 5.5 Parallel rays emerging from a concave lens.

It must lie on the opposite side of the perpendicular from CE. In addition, the inclination of the refracted ray EG must be larger than that of the incident ray CE, since the light is going from glass to air. Thus, EG diverges from FH, which is parallel to CE. Similarly, DB diverges from CA.

We have shown that when the parallel rays inside the concave lens are refracted, divergent rays are produced at both surfaces. Now, one may reinterpret the diagram as describing light paths $ACEG$ and $BDFH$. From this point of view, convergent rays AC and BD are refracted by the concave lens to yield divergent rays EG and FH on the opposite side of the lens. In the case considered, convergent rays are actually made divergent by a concave lens. In general, the effect of a concave lens may be less dramatic, merely diminishing the convergence of convergent rays or increasing the divergence of divergent rays.

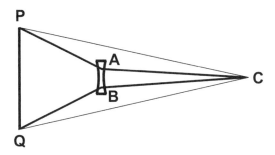

Figure 5.6 Minification by a concave lens.

We now show that concave lenses make objects appear smaller. Figure 5.6 depicts an object PQ viewed by an eye at C through a concave lens. Rays from opposite edges of the object converge toward the concave lens and are made less convergent. The angle subtended by the object when viewed through the lens is ACB. Clearly, angle ACB is smaller than angle PCQ, which the object would subtend if the lens were not present. This reduction of the angle subtended at the eye by the object means that the object appears smaller.

Eye glasses

Eye glasses may consist of either convex or concave lenses. Recall that the convergence points are closer to a convex lens for distant objects than for nearby objects. When a nearsighted eye views a distant object, the lens of the eye cannot accommodate sufficiently to shift the convergence points all the way back to the retina. After maximal accommodation, the rays emerging from the lens of the eye remain too convergent.

If one could put a concave lens behind the lens of the eye, these convergent rays could be made less convergent, and the convergence points could be moved back to the retina. It turns out that the same effect is obtained by putting the concave lens in front of the eye. Thus, glasses for nearsighted people utilize concave lenses.

Conversely, the convergence points are farther from a convex lens for nearby objects than for distant objects. When a farsighted eye views a nearby object, the convergence points lie behind the retina. The lens cannot accommodate sufficiently to move them forward to the retina. In effect, the rays emerging from the lens of the eye remain insufficiently convergent. Thus, glasses for farsighted people utilize convex lenses to make these rays more convergent.

It is not difficult to tell if a pair of glasses is made with convex or concave lenses. View an object through the glasses while holding them close to your face. Then, gradually increase the distance between the glasses and your face. If the object grows in size, the lenses are convex, whereas if the object diminishes in size, the lenses are concave.

The Galilean telescope

The Galilean telescope, like the Keplerian, achieves magnification by allowing the user to view objects through a convex lens from a position close to the convergence points. Again, the challenge is to compensate for the blurring that occurs in that vicinity. Galileo's design places a concave lens between the convex lens and the convergence points to render the convergent rays slightly divergent. By choosing a concave lens that produces only slightly divergent rays, blurring due to excessive divergence is also avoided.

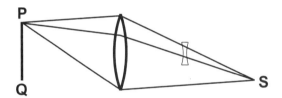

Figure 5.7 Layout of the Galilean telescope.

The layout of the Galilean telescope is shown in Figure 5.7. Rays from P diverge against the front surface of the objective and, in the absence of the ocular lens, their refractions would converge at S. In the figure, the position of the ocular is shown, but its effect on the rays is not indicated. Note that only a portion of the rays diverging from P against the objective are refracted toward the ocular.

Figure 5.8 traces rays from both P and Q and shows how they are refracted by the ocular into the pupil of the eye. The refracted rays from each object point are convergent as they proceed from the back surface of the objective to the front surface of the ocular. The concave ocular makes these slightly divergent and suitable for viewing.

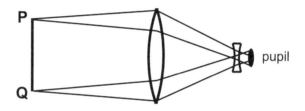

Figure 5.8 Operation of the Galilean telescope.

Note that rays from P, which is above the axis, enter the pupil from above the axis, and those from Q enter from below. Thus, the Galilean telescope shows objects in their true orientation. Finally, magnification by the Galilean telescope is explained just as for the Keplerian telescope. We compare the angle subtended by the object at the eye with and without the telescope. The effect of the telescope is to increase the angle between rays from opposite edges of the object.

Sine law of refraction

Recall that Kepler's two rules of refraction gave only an approximate account of the relationship between the inclination of the incident ray and the angle of refraction. The agreement with experimental data was excellent for small inclinations, but not for larger angles. Modern optics gives a mathematical description of refraction that is in agreement with the data for any angle. This is the sine law of refraction.

Interestingly, there is evidence that the Arabic scholar Ibn Sahl had some awareness of the sine law in the tenth century, but it was unknown in Europe when Kepler was active in optics. The sine law was rediscovered by several European scientists within a few decades after *Dioptrice* was published. Those credited with the discovery include Thomas Harriot, Willebrord Snell, and René Descartes.

For light passing from one transparent medium to another, the sine law of refraction is stated as

$$n_1 sin\theta = n_2 sin\phi,$$

where n_1 and n_2 are the *indices of refraction* of the two transparent media. In *Dioptrice*, Kepler deals uniformly with refraction at the interface of air and glass. Thus, the indices of refraction are always those of air and glass and are not explicitly stated. Note that the naming of angles in relation to the sine law differs from that used by Kepler. As illustrated in Figure 5.9, the inclination of the incident ray is today called the *angle of incidence*. What Kepler would have called the inclination of the refracted ray is today called the *angle of refraction*.

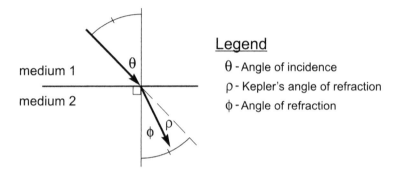

Figure 5.9 Sine law of refraction.

The index of refraction of air is very close to 1.0 and that of glass is close to 1.5. Thus, for air and glass, the sine law of refraction becomes

$$sin\theta = \frac{3}{2}sin\phi.$$

Kepler's approximate rule of refraction is tantamount to the small-angle approximation in which the sines are replaced by the angles themselves:

$$\theta = \frac{3}{2}\phi.$$

Although this equation is not obviously equivalent to Kepler's rules of refraction, it is a simple exercise to derive his rules from it. We do this for the case of light entering glass from air. From Figure 5.9, we have $\rho = \theta - \phi$ and, from this, $\phi = \theta - \rho$. We do a bit of algebra:

$$\theta = 3\phi/2$$
$$\theta = 3(\theta - \rho)/2$$
$$3\rho/2 = 3\theta/2 - \theta$$
$$3\rho = \theta$$
$$\rho = \theta/3,$$

The last line, $\rho = \theta/3$, states that the angle of refraction is one-third of the inclination of the incident ray.

Table 5.1 compares the angle of refraction given by Kepler's approximate rule to that given by the sine law. For inclinations less than $10°$, it is seen that Kepler's rule is accurate to better than 1%. Even at angles as large as $30°$, the error is only 5%, which may be tolerable in many situations.

The rediscovery of the sine law of refraction was fortunate, in that it enabled subsequent generations to carry out more exact treatments of optical systems than had been possible for Kepler. But Kepler entered the scene when virtually nothing had been done. His approximate rule for refraction was perfectly adequate for understanding the basic function of lenses and their use in a telescope. Does the exact sine law of refraction

Table 5.1 Refraction ρ versus inclination θ.

θ (degrees)	ρ (degrees)		Error (%)
	Sine law	Kepler's rule	
0	0.00	0.00	0.0
10	3.35	3.33	0.6
20	6.82	6.67	2.2
30	10.53	10.00	5.0

render *Dioptrice* obsolete? It does not. The availability of the sine law encourages blind calculation, whereas *Dioptrice* helps us develop an intuition for optics.

Appendix A

Comparing angles

Adding angles

Angles are represented in diagrams as two line segments sharing an endpoint. The common endpoint is called a *vertex*. We will call the line segments *sides* of the angle. One side will be called the *initial side* and the other the *terminal side*. In what follows, we will assume that an angle can be rotated or moved from one location to another without changing anything essential to the angle.

Angles are added by placing them adjacent to one another so that the vertices coincide and the terminal side of the first angle coincides with the initial side of the second. The resultant angle has as its initial side the initial side of the first angle and as its terminal side the terminal side of the second angle. Thus, in Figure A.1, angles ABC and CBD are added to obtain angle ABD.

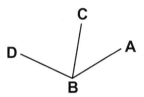

Figure A.1 Adding two angles.

Angles are equal if, when they are positioned with the vertices and initial sides coinciding, the terminal sides also coincide. In Figure A.2 angles ABC and DBE are equal. Note that there is no requirement that the sides of the angles have equal lengths. In fact, the lengths of the sides are completely irrelevant.

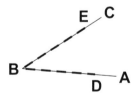

Figure A.2 Equality of angles.

Supplementary angles

Angles ABC and CBD in Figure A.3 are *supplementary angles*, or *supplements* of one another. Angles are supplementary when their sum is an angle whose initial and terminal sides lie on a single straight line, as do sides AB and BD of angle ABD in the figure.

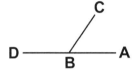

Figure A.3 Supplementary angles.

It is true that angle ABD doesn't look much like what we normally think of as an angle. But if we accept the idea that the sum of two angles is always another angle, then here we must accept ABD as an angle. For lack of a better term, we'll refer to this angle as a *straight angle*.

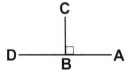

Figure A.4 A straight angle equals two right angles.

When two supplementary angles are equal, they are both half of a straight angle, and each is called a *right angle.* The sides of a right angle are said to be *perpendicular* to one another. Right angles are indicated in drawings by placing a small square at the corner where the two sides meet, as illustrated in Figure A.4.

In Figure A.5 two line segments cross, resulting in four angles. To keep the figure uncluttered, the angles are designated by single Greek letters α (alpha), β (beta), γ (gamma), and δ (delta). Note that there are four pairs of supplementary angles in the figure.

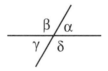

Figure A.5 Supplements of the same angle are equal.

By considering both pairs α and β and β and γ, we will be forced to conclude that angles α and γ are equal. First, start with α and add β. The sum is two right angles. Next, start with γ and add β. The sum is again two right angles. In both cases, we started with an angle and added β to obtain an angle equal to two right angles. It follows that what we started with in both cases must have been the same thing. That is, α equals γ. Using the same argument, we can show that angles β and δ in the figure are also equal.

Conclusion: If two angles are supplements of the same angle, they are equal.

Parallels cut by a transversal

Lines, as opposed to line segments, are defined as continuing indefinitely in both directions. In Figure A.6 lines 1 and 2 are transected by line 3, called a *transversal*. Because the labeled angles lie between the top and bottom lines, they are referred to as the *internal angles*. The unlabeled angles outside the lines are called the *external angles*. Angles α and β at the top are supplementary. Their sum is two right angles. Similarly the sum of γ and δ at the bottom is two right angles. Thus, the sum of the four internal angles is four right angles.

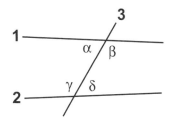

Figure A.6 Two lines cut by a transversal.

Next, we compare the sum of the two internal angles on the right side of the transversal with the sum of the two on the left side. If the sums are equal, then each sum must be two right angles, since the sum of all the internal angles is four right angles. Otherwise, one sum is greater than two right angles and the other is less than two right angles.

This brings us to Euclid's famous fifth postulate about parallel lines, which states that two lines cut by a transversal are parallel if and only if the sums of the internal angles to the right and to the left of the transversal both equal two right angles. The postulate goes on to say that if the sum is less than two right angles on one side, then the lines intersect on that side, and consequently are not parallel.

Starting with Euclid's postulate, we can establish a very useful result. Assume that lines 1 and 2 in Figure A.6 above are in fact parallel. If so, the sum of α and γ is two right angles. But angles α and β are supplementary, so their sum is also two right angles. Thus, if we start with γ and add α, we end up with the same result as when we start with β and add α. This can only be true if γ and β are equal. In the same way, one can show that α and δ are equal.

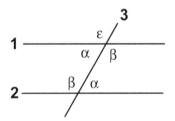

Figure A.7 Parallel lines cut by a transversal.

The labeling of angles in Figure A.7 illustrates this relationship among the internal angles formed by two parallel lines and a transversal. We have also labeled one of the external angles ϵ (epsilon). Now, angles ϵ and β are both supplements of α, so ϵ equals β. Continuing in this way, one can show that the eight angles formed by a transversal and two parallels come only in two sizes, one of which is the supplement of the other.

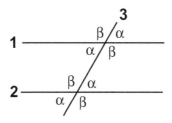

Figure A.8 Angles formed by a transversal cutting two parallels.

Figure A.8 summarizes all the relationships we have demonstrated among the angles formed by a transversal and two parallels. What is important to remember is that the angles are laid out identically around the two intersections.

Appendix B

Measuring angles

Fractional angle, arc length

Appendix A discussed the equality of angles or sums of angles, but never mentioned the numeric size of an angle. In fact, there are several different ways of assigning numeric sizes to angles. Most commonly, angles are measured in degrees. This system of measurement depends on a relation between angles and circles.

Figure B.1 shows a circle drawn with its center at the vertex B of angle ABC. Note that the size of the circle is arbitrary, and is determined solely by convenience. Consider the portion of the circle that lies between the sides of the angle, proceeding counterclockwise from A to C. Clearly, the larger the angle, the larger will be the fraction of the circle inside the angle.

Thus, the size of an angle could be defined as that fraction of the circle that lies inside the angle. We will call this measure of an angle's size the *fractional arc length*. Consider a right angle, which marks off one-quarter of the circumference of a circle. Its fractional arc length would be $1/4$, or carrying out the division indicated by the fraction, 0.25.

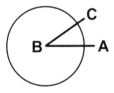

Figure B.1 Angle and arc length.

Finding the fractional arc length of an angle requires two separate measurements and a calculation. First, we measure the length of the arc between the initial and the terminal sides of the angle. Next, we either measure the circumference of the circle directly, or measure the radius or diameter of the circle and calculate the circumference. Finally, we divide the arc length by the circumference.

The familiar measurement of angles in degrees is closely related to the fractional arc length. To find the size of an angle in degrees, we simply take the fractional arc length and multiply it by 360. This procedure can be expressed by an algebraic formula.

Let θ be the angle of interest, s the length of the arc inside the angle, and r the radius of the circle. Then, since the circumference of the circle is $2\pi r$,

$$\theta = 360\frac{s}{2\pi r}. \tag{B.1}$$

In practice, part of this calculation is usually carried out in advance when a *graduated circle* is manufactured. A graduated circle is any rigid circular structure, the circumference of which has been divided into small equal intervals. The protractor, known to every student of geometry, is a familiar example. But graduated circles have also been incorporated into a large variety of measuring instruments.

Graduated circles are all used in essentially the same way. We position the center of the circle at the vertex of the angle to be measured. Then we note the number of graduations that lie between the sides of the angle. This procedure is equivalent to determining the fractional length of the arc because each subdivision on the graduated circle is a predetermined fraction of the circle's circumference.

In navigation, surveying, and astronomy, it is necessary to measure angles to a high degree of precision. Degrees are subdivided into 60 *minutes*, and minutes are subdivided into 60 *seconds*. If angles are to be measured using a graduated circle, an enormous number of equally spaced marks must be accurately placed around the periphery. Fabricating a good graduated circle is not a trivial undertaking. However, if we are willing to tolerate a little more mathematical complexity, then we can measure angles without a graduated circle. Kepler was a better mathematician than craftsman, so that was his approach.

Tangent of an angle

Figure B.2 depicts an angle θ with initial side AB, vertex B, and terminal side BC. A circle drawn with its center at B intersects the initial side AB at E. Since it is a radius of the circle, the line segment BE is perpendicular to the circle at E.

Line LN is what is called a *tangent* to the circle. It touches the circle at E but does not pass into the interior of the circle. For this to be true, LN must be perpendicular to BE. What we are going to be interested in is DE, the portion of the tangent line marked off by the initial and terminal sides of θ.

Keeping the initial side of the angle AB fixed, imagine pivoting the terminal side BC at the vertex, so as to increase or decrease the size of the angle. The intersection D of the terminal side and the tangent line rises or falls as the angle increases or decreases. Thus, the portion of the tangent line marked off by the sides of the angle is a measure of how large the angle is.

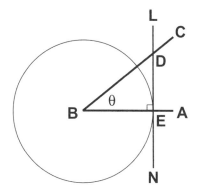

Figure B.2 Circle and tangent line.

As it stands, however, this measure of the angle depends on the size of the circle. For example, if the circle were made larger by a factor of 2, then the same angle θ would mark off a length along the tangent line that is twice as long. But this is easily fixed.

The radius is a convenient index of the circle's size. Thus, we can compensate for the size of the circle by dividing the length along the tangent line by the radius. The result is the tangent of the angle, abbreviated $\tan\theta$. In the present example, $\tan\theta = DE/BE$. Obviously the tangent of an angle, which is a number, differs from the tangent line, but they are related.

In practice, we usually find the tangent of an angle as the ratio of two measured lengths, and then use a conversion to find the corresponding angle

Table B.1 Tangents.

tan θ	θ (degrees)
0.000	0
0.087	5
0.268	15
0.577	30
1.000	45
1.732	60

expressed in degrees. Table B.1 lists a few tangents with their corresponding angles. Of course, a conversion would require many more entries than this to be useful. Moreover, in our time, the conversion is usually carried out by means of an electronic calculator.

Sine of an angle

The sine of an angle θ, abbreviated $sin\theta$, is yet another measure of an angle's size. The sine and the tangent are really two variations on the same theme. Figure B.3 depicts a second line JK that, like the tangent line DE, is perpendicular to the initial side of angle θ. Now, we can draw many lines that happen to be perpendicular to the initial side of θ. What distinguishes the tangent line is that it is drawn through point E where the initial side of the angle intersects the circle. What distinguishes line JK is that it is drawn through point J where the terminal side intersects the circle. Note that BE and BJ are both one radius of the circle in length.

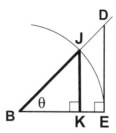

Figure B.3 Sine of an angle.

Now, just as the span DE of θ along the tangent line increases or decreases as θ does, so too the span JK along the perpendicular through J increases or decreases with θ. As we did with the tangent, we divide JK by the radius to construct a measure of the angle that is independent of the size of the circle. Thus, $sin\theta = JK/JB$.

Tangent, sine, right triangle

In this section we will be concerned with ratios, or comparisons of two numbers. We will regard ratios as essentially the same as fractions, and we will think of fractions as implying a division of the numerator by the denominator. A standard way of representing a ratio is to form a fraction with the first number in the numerator and the second in the denominator. Then, the result of dividing the numerator by the denominator indicates how the first number compares to the second.

If the result is greater than 1.0, the first number is larger than the second. The more the result exceeds 1.0, the larger the first number is in comparison to the second. Similarly, if the result is between 0.0 and 1.0, then the first number is smaller than the second. The closer the result is to 0, the smaller the first is in comparison to the second.

Now, there is a way of defining the tangent and sine that is often more convenient in practice than the one we presented above. The new definitions rely on a property of triangles discussed in the last section of Appendix C:

- Specification of two of the three angles of a triangle suffices to fix the relative lengths of the triangle's sides.

As an application of this idea, suppose we construct a triangle, one angle of which is θ and another of which is a right angle. Having specified these two angles, the ratios of the sides will be fixed, irrespective of how large the triangle is. We will define one of these ratios as the tangent of θ and another as the sine of θ. Figure B.4 depicts such a triangle. The tangent of θ is the ratio CB/AB, and the sine of θ is the ratio CB/CA.

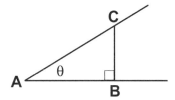

Figure B.4 A right triangle constructed from angle θ.

Now, these definitions would be arbitrary and difficult to remember, were it not for the preceding discussion of the tangent and sine in the context of a tangent line to a circle. As before, it is the line segment opposite θ that provides a measure of how large θ is. The larger θ is, the longer CB must be to span the angle.

Previously, we divided the length of the line segment spanning θ by the radius of the circle, so as to factor out the size of the circle. In a similar way, we now form a fraction consisting of the length of the side opposite θ divided by the length of one of the other sides. This measure of θ is independent of the size of the triangle because of the fact that two angles, in our case θ and the right angle, suffice to determine the relative lengths of the sides.

By now it is obvious that side CB opposite θ belongs in the numerator. To remember that it is the second side of the right angle that goes in the denominator for the tangent, recall that the tangent line intersects a circle at right angles to the radius.

To remember that it is the hypotenuse that belongs in the denominator of the sine, think of the word *sinus*. A sinus is a cavity, and the hypotenuse joins the side opposite θ at an angle of less than $90°$, forming a sort of recess or cavity.

Finally, look back at Figure B.3, where we identified the sine of θ as JK/JB. The tangent of θ in Figure B.3 is DE/DB, just as it was in Figure B.2. But the ratios of corresponding sides in two different triangles are the same if the angles are the same. Therefore, since triangles BDE and BJK have equal angles, $JK/BK = DE/DB$, and it is also true that $tan\theta = JK/BK$. Similarly, an alternative expression for the sine is $sin\theta = DE/DB$.

Sine of an obtuse angle

An angle smaller than $90°$ is called *acute*, while any angle larger than $90°$ is called *obtuse*. We need to give some thought to how the sine of an obtuse angle can be defined. In the case of an acute angle, we had two ways of thinking about the sine. One definition was in terms of a right triangle. In Figure B.5 triangle ABC associated with the obtuse angle θ is not a right triangle. Consequently, there is no obvious definition of the sine in terms of a right triangle.

Fortunately, our other approach to defining the sine does work for an obtuse angle. Again, we draw

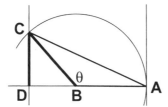

Figure B.5 Sine of an obtuse angle.

the tangent line to the circle at A and we also draw
a parallel to the tangent through point C, where the
terminal side of the angle intersects the circle. This
parallel does not intersect the initial side of the angle,
as in the case of an acute angle, but it does form an
intersection D with the diameter of the circle upon
which the initial side lies. We take the distance CD
between these two intersections as the numerator of
the fraction we will form. As usual, the denominator
is the radius of the circle. Thus, we define the sine of
the obtuse angle θ as CD/CB.

Although the triangle ABC that contains θ is not a
right triangle, we have formed a second triangle BCD
that is. In fact, the sine of angle CBD in this triangle is
equal to CD/CB, which is precisely what we decided
$sin\theta$ should be. Now, these angles θ and CBD, which
have the same sine, are supplements of one another.
Do supplementary angles always have the same sine?
The way we have defined the sine of an obtuse angle
guarantees that they do.

Conclusion: The sine of an angle equals the sine of its supplement.

Rule of sines

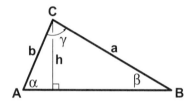

Figure B.6 Rule of sines.

The rule of sines is a relation involving the angles of a triangle and the lengths of its sides. Observe in Figure B.6 that $sin\alpha = h/b$ and $sin\beta = h/a$. Note that both expressions involve h, the perpendicular distance of point C from side AB. Solve both equations for h and equate the results to find that $h = b(sin\alpha) = a(sin\beta)$. Then $(sin\alpha)/a = (sin\beta)/b$. The same argument works pairing angle α with γ or β with γ. The result is the following very useful rule of sines:

$$\frac{sin\alpha}{a} = \frac{sin\beta}{b} = \frac{sin\gamma}{c}$$

Sine and tangent of small angles

Figure B.7 is a magnified view of a very small angle θ. The sides of the angle are only partially shown. They continue to the left of the figure until they intersect at the vertex of the angle. The arc s, which spans the angle, lies on a circle of radius r whose center is at the vertex.

The figure also shows the line segments that we used above to define the tangent and sine of θ. The labels a and b represent the lengths of the line segments. We obtain $tan\theta$ and $sin\theta$, respectively, when these are divided by r.

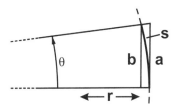

Figure B.7 Sine and tangent of a small angle.

Now, as θ is made progressively smaller, the line segments a and b draw ever closer to one another, while the arc s is "squashed flat" between them. Thus, for small angles, the arc length and the lengths of the two line segments will be approximately equal. The smaller the angle, the more nearly equal they are.

We draw several conclusions. First, the lengths of these line segments yield the tangent and sine when divided by the radius of the circle. Therefore, the smaller the angle, and the more nearly equal the line segments are, the more nearly equal the tangent and sine are.

At the same time, when the angle is small, the lengths of the line segments a and b are very nearly equal to the arc length s. Therefore, we can obtain a good estimate of the arc length by multiplying either $tan\theta$ or $sin\theta$ by r:

$$r tan\theta \approx s, \qquad\qquad \text{(B.2)}$$
$$r sin\theta \approx s. \qquad\qquad \text{(B.3)}$$

Again, the smaller θ is, the better the approximation.

Finally, when the angle is expressed in degrees, we can solve equation B.1 from the first section of this appendix to obtain

$$s = \frac{2\pi r}{360}\theta.$$

Substituting the expression for arc length s into equations B.2 and B.3 and canceling the radius, we obtain

$$tan\theta \approx \frac{2\pi}{360}\theta,$$

$$sin\theta \approx \frac{2\pi}{360}\theta.$$

Thus, for small angles, the tangent and sine are approximately proportional to the angle, and the constant of proportionality is $2\pi/360$ in both cases.

Table B.2 Tangents and sines.

θ (degrees)	$\frac{2\pi}{360}\theta$	$tan\theta$	$sin\theta$
0	0.000	0.000	0.000
5	0.087	0.087	0.087
15	0.262	0.268	0.259
30	0.524	0.577	0.500
45	0.785	1.000	0.707

To illustrate how the approximation depends on the size of θ, we list in Table B.2 values for $tan\theta$ and $sin\theta$ as the angle θ varies from $0°$ to $45°$. For angles up to $15°$, the tangent and sine are proportional to the angle within a few percent.

Appendix C

Triangles

In Appendix B we claimed that the angles of a triangle determine the relative lengths of its sides. This appendix validates that claim. Note, however, that none of the material in this appendix is used directly by Kepler in *Dioptrice*.

The angles of a triangle

One of the most commonly used facts of plane geometry is that the angles of any triangle add up to $180°$. This is easily demonstrated using the construction of two parallels cut by a transversal, which was discussed at the end of Appendix A. Figure C.1 shows two identical but otherwise arbitrary triangles, ABC and CDE. They are positioned so that the two triangles share vertex C. In addition, the triangles are oriented so that sides AC and CE lie along the same straight line.

The angles of both triangles are α, β, and γ. The critical observation is that the three angles, γ, δ, and α, which all have the point C as their vertex, occupy the entire space above line ACE. Consequently, their sum is two right angles, or $180°$.

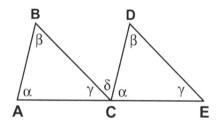

Figure C.1 Angles of a triangle.

We are going to prove that the sum of α, β, and γ is equal to that of α, δ, and γ by showing that the angle β is equal to δ. Since we have assumed no special property regarding the triangles ABC and CDE, it will follow that the three angles of any triangle have a sum equal to $180°$.

Here is the proof. The line segments AB and CD are parallel to one another because they are both inclined by the same angle α to the straight line ACE. Then BC is a transversal cutting these two parallel lines.

Recall that all of the angles formed by two parallels and a transversal come in only two sizes, depending on their position. They are either some angle θ or its supplement, $180 - \theta$. Now, reference to Figure C.2 will show that angles β and δ in Figure C.1 are positioned so as to be equal. We conclude that the three angles of any triangle have a sum equal to $180°$.

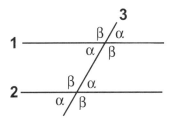

Figure C.2 Angles formed by a transversal cutting two parallels.

There is a useful corollary to this. Suppose that two angles α and β, having fixed sizes, are both present in two different triangles. Then the third angle must be the same in each triangle. This is because the sum of the first two angles is the same in each triangle, and the third angle must in each case bring the total to $180°$.

The area of a triangle

This section derives and generalizes the well-known formula for the area of a triangle. Then we use the formula to obtain a result needed in the next section.

The area of a right triangle is readily found, as illustrated by Figure C.3. Starting with the right triangle ABC, an identical copy ABD is made and the two triangles are joined so as to form a rectangle. The area of the rectangle is ab, where a and b are the lengths of its sides. Since the diagonal divides the rectangle into two equal triangles, the area of each triangle is $ab/2$.

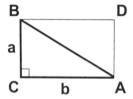

Figure C.3 Area of a right triangle.

Two rectangles are required to find the area of a triangle if none of its angles is a right angle. When the angles are all acute, it is possible to start with any side of the triangle and draw a line perpendicular to it that passes through the opposite vertex. An example is given in Figure C.4.

In the figure, BD is perpendicular to AC. The length of the perpendicular is a, and the parts of AC on opposite sides of D have lengths b and c. The perpendicular divides triangle ABC into two right triangles of areas $ab/2$ and $ac/2$. So the area of triangle ABC is the sum $ab/2 + ac/2 = a(b + c)/2$. Since $b + c$

is the length of side AC, the area of the triangle is the length of side AC times the *perpendicular distance* of B from AC divided by two.

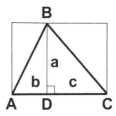

Figure C.4 Area of a triangle from the sum of two triangles.

Let's compare the rules for the area of a right triangle and a triangle with three acute angles:

- Right triangle: Multiply the lengths of the two perpendicular sides and divide by two.

- Three acute angles: Multiply the length of one side by the perpendicular distance to the opposite vertex and divide by two.

When a triangle contains an obtuse angle, the sides opposite the acute angles become troublesome. It is impossible to draw perpendiculars from these sides that pass though the opposite vertex. Figure C.5 shows how this can be handled. We extend a line coinciding with side AC to the left of A. Then a perpendicular from D passes through B. Although BD makes no contact with AC, we call it the perpendicular distance of B from AC because it is the shortest distance from B to the line that AC is on.

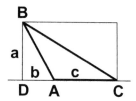

Figure C.5 Area as the difference of two triangles.

In Figure C.5 the area of triangle ABC may be found as the area of triangle DBC minus the area of triangle DBA. The area of DBC is $a(b+c)/2$ and the area of BDA is $ab/2$. Therefore, the area of ABC is the difference $a(b+c)/2 - ab/2$, which equals $ac/2$.

So the area of a triangle with an obtuse angle obeys the same rule as our previous examples. The area is still the length of one side times the perpendicular distance from that side to the opposite vertex, divided by two. Thus, by generalizing the concept of perpendicular distance, we have found a rule for the area of a triangle that applies in all cases.

Conclusion: The area of any triangle is the length of any of its sides times the perpendicular distance from that side to the opposite vertex, divided by two.

We use this rule to derive a result needed in the next section. In Figure C.6 triangles ABC and DEF are drawn between parallel lines L_1 and L_2. Each triangle has one side on L_1 and the opposite vertex on L_2.

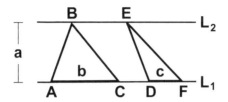

Figure C.6 Triangles drawn between two parallels.

The distance between the parallels is a and the lengths of sides AC and DF on L_1 are b and c, respectively. The area of triangle ABC is $ab/2$ and that of DEF is $ac/2$. Now, the ratio of area ABC to DEF is $ab/2$ divided by $ac/2$, which equals b/c. But this is precisely the ratio of the lengths of AC and DF.

We will use two variations of this result in the next section. In the first instance, two triangles with a side on L_1 share a single point as their vertex on L_2. But this doesn't affect the perpendicular distance from the vertex to the sides on L_1, so the ratio of areas remains equal to the ratio of the sides. In the second case, the vertices on L_2 are distinct but the triangles share a common side on L_1. Here, since the sides on L_1 have equal lengths, the triangles have equal areas.

The sides of a triangle

In this section, we establish that the angles of a triangle determine the relative lengths of the sides. Figure C.7 depicts two triangles, with identical angles α, β, and γ. Given only that the angles are the same, we will show that the ratio of lengths of the left and right sides is the same in each triangle. Similar demonstrations would establish that the ratios of left to bottom and right to bottom are also the same.

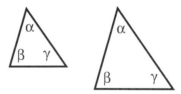

Figure C.7 Two triangles with the same angles.

In Figure C.8 the two triangles have been super-imposed so that their angles α coincide. The left and right sides of the smaller triangle have lengths a and b, respectively, and the corresponding sides of the larger triangle have lengths e and f. The longer sides exceed the shorter ones by lengths c and d. What we want to show is that $e/f = a/b$.

Proceeding to Figure C.9, the vertices are labeled and diagonals AE and CD have been added. All other labels have been omitted to avoid clutter. Our first task is to establish that sides DE and AC are parallel.

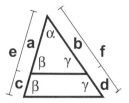

Figure C.8 Lengths of the sides.

DE and AC are cut by the transversal BA. Angle ADE is the supplement of BDE, and because BDE and DAC are equal, ADE is also the supplement of DAC. Thus, the sum of ADE and DAC is two right angles, and by Euclid's fifth postulate DE is parallel to AC.[1]

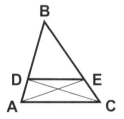

Figure C.9 Ratios of areas and sides.

Next, we infer three relations from Figure C.9, based on the discussion of Figure C.6. First, the ratio of areas ADE and BDE equals the ratio of lengths DA and BD because DA and BD are on the same line and

---

[1]The method of proof here is essentially that of Proposition 2 in Book 6 of Euclid's *Elements*.

triangles ADE and BDE share vertex E. Similarly, the ratio of areas CDE and BDE equals the ratio of lengths EC and BE. Finally, the areas of triangles ADE and CDE are the same, because they are drawn between two parallel lines and share the side DE.

Now, since $ADE = CDE$, we have $ADE/BDE = CDE/BDE$. But the ratios ADE/BDE and CDE/BDE equal DA/BD and EC/BE, respectively. Therefore, $DA/BD = EC/BE$ or, in the notation of Figure C.8, $c/a = d/b$.

We finish the proof with a little bit of algebra. Starting with $c/a = d/b$, we add 1 to both sides of the equation using the trick of representing 1 by a/a on the left and by b/b on the right:

$$c/a + 1 = d/b + 1$$
$$c/a + a/a = d/b + b/b$$
$$(c + a)/a = (d + b)/b$$

From Figure C.8, we see that $c + a = e$ and $d + b = f$. Making these substitutions and cross-multiplying, we obtain

$$e/a = f/b$$
$$e/f = a/b,$$

which is what we wanted to prove.

References

[1] Caspar, Max. *Kepler.* New York: Dover Publications, 1993. An authoritative biography by one of the editors of Kepler's collected works.

[2] Kepler, Johannes. *Dioptrice.* In *Johannes Kepler Gesammelte Werke*, Vol. 4, pp. 355–414. C. H. Beck, 1937. Kepler's collected work, in the original languages. *Dioptrice* appears in Latin.

[3] Kepler, Johannes. *Optics: Paralipomena to Witelo and Optical Part of Astronomy.* Translated by William H. Donahue. Santa Fe: Green Lion Press, 2000. Kepler's first and major work on optics, translated by the preeminent Kepler scholar of our time.

[4] Lindberg, David C. *Theories of Vision from Al-Kindi to Kepler.* Chicago: The University of Chicago Press, 1976. A fascinating account of the quest to understand vision.

[5] Malet, Antoni. "Kepler and the Telescope," *Annals of Science*, Vol. 60, pp. 107–136. Taylor & Francis, 2003. An overview of *Dioptrice* and discussion of its significance by an expert on seventeenth-century science and mathematics.

[6] Voelkel, James R. *Johannes Kepler and the New Astronomy*. Oxford: Oxford University Press, 1999. An excellent introduction to Kepler's life and work.

[7] Watson, Fred. *Stargazer: The Life and Times of the Telescope*. Cambridge: Da Capo Press, 2005. A very readable history of the telescope.

Index